Stress Relief Coloring Book
For Adults

Copyright 2016
Dean R. Giles

All rights reserved. No part of this book or coloring pages may be reproduced or used in any form, whether electronic or mechanical, without written prior consent.

What is a Mandala?

The word Mandala comes from India. Its translated meaning is "circle". Originally, mandala had a kind of comprehensive meaning as a representation of the cosmos or everything. Historically it was used in religious art from Christianity through Buddhism, and was found in diverse cultures from Tibet to the ancient Aztecs. The concentric circle pattern has been recognized as aesthetically pleasing and comforting. Mandalas have become popular because of these aesthetic qualities. The mandalas produced here are designed to provide hours of relaxation and stress relief while entertaining the mind and the visual senses.

Why a Stress Relief Coloring Book

Many adults have heard of the stress relieving attributes of coloring, but after purchasing a coloring book found great stress in just getting started. It all looks quite overwhelming. My first suggestion is to just take one deep breath, then let it out slowly. Keep in mind that it is just coloring! Coloring is something you knew how to do at four or five years old, and like riding a bike, you will remember how very quickly. The important part is to remember that there is no right or wrong way to do the coloring, let your intuition, mood, or just the moment help your decide on what colors to use, whether to choose specific shapes to color the same or similar colors, or whether to group the layers or sections with the colors or color groups you want. Strive for some level of harmony, decide on a strategy up front, but by all means change the strategy as you go when you feel prompted. The point of the activity is to just let it flow, and to feel fulfilled creatively as it comes to an end.

How Do I Select What to Color With?

Well, you have matured since those early years of coloring, crayons probably won't give you the creative rush you are looking for. Select colored pencils, fine tipped markers, or colored pens. Usually paying a little more for a better tip will pay dividends in the end when it comes to the look of the finished product and the ability to fill in the tiny spaces.

How Do I Choose What Colors to Use?

Color can be a personal thing, but my experience is that detailed pictures can use many colors. Don't be afraid to, "paint with all the colors of the wind." Experiment, you will find colors and blends that appeal to you more than others.

When it comes to selectin a strategy for what colors go together you can do one of the following:
Just make it up—you know, close your eyes and grab a pencil.
Use a color wheel—many online versions exist, and they can help you out.
My favorite, select a painting you like and mimic the colors you find in it. You will be surprised what you find out about the colors that go together by the pallet used by other artists.

Most importantly, don't stress about it. A rainbow has many colors in it but seems to harmonize perfectly. As you color you will likely discover what colors and color schemes fill you with the most awe and inspiration. Let it flow as you go, and don't let anything trouble you.

How to Use This Book

Detailed mandalas, whether coloring, or simply meditating on the shape and repeating patterns, can be stress relieving. It is my hope that you can enjoy the pictures and creations exhibited here with the intent to add a little stress relief, peace, and tranquility to your life, today and beyond.

Enjoy the journey. The end is uncertain, but the journey can be an endless adventure. Make the best of your adventure.

Dean R. Giles
dean@austinsgift.com
FreeColoringBook.org

Congratulations on your purchase of this coloring book and art gallery. As a purchaser of this book, you are entitled to a Free Gift. Please register to accept this valuable gift, a free PDF coloring book found at http://www.FreeColoringBook.org/gift .

Thank you for purchasing this coloring book and art gallery. If you enjoyed the book, please leave a review. Reviews are so needed and appreciated by Independent Authors.

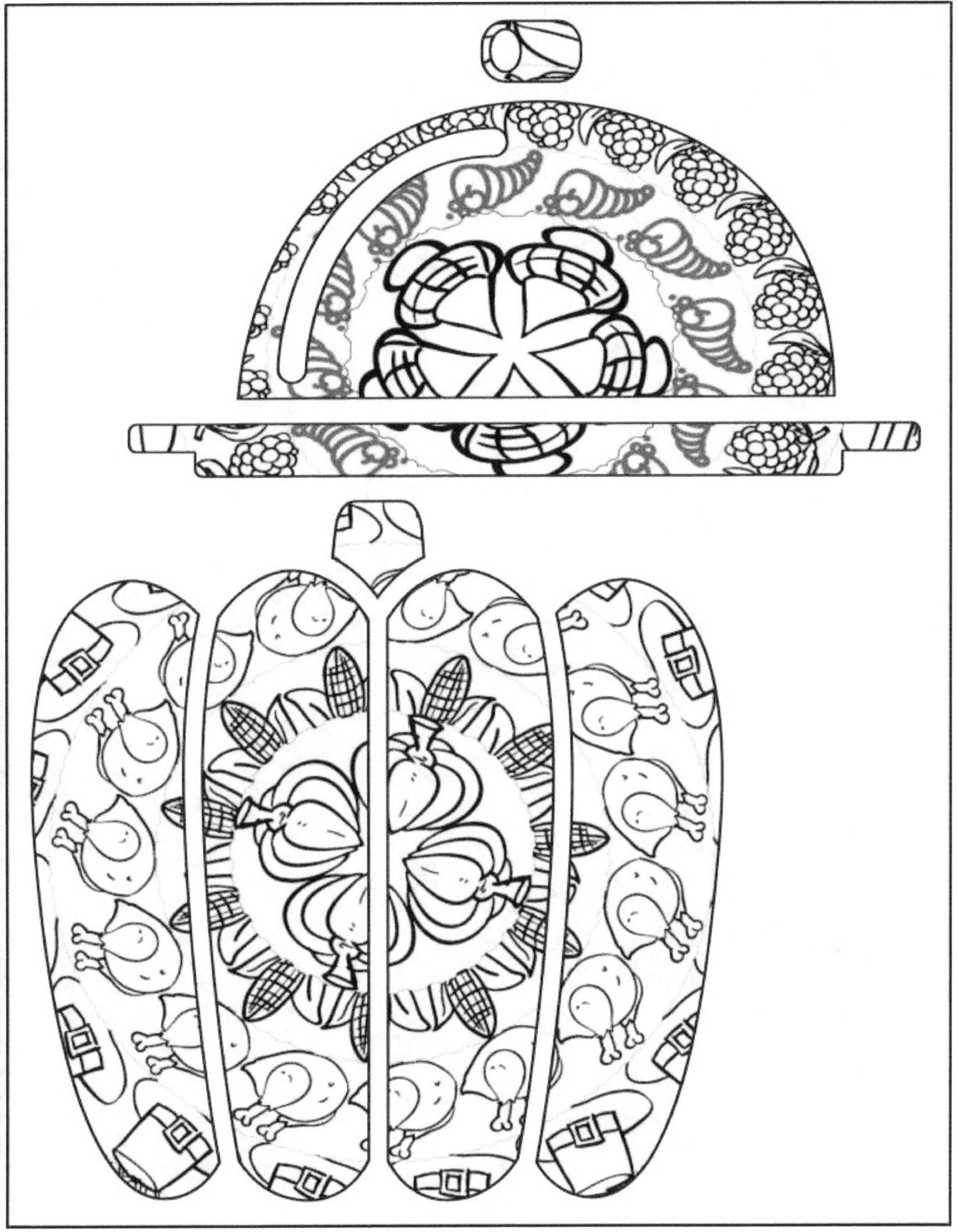

Did you enjoy this coloring book? If so, I have a special gift for you, our valued customer.

Get Your Free Gift!

Please register to download your free gift at **FreeColoringBook.org/gift**.

Looking for more coloring adventures?

http://amzn.to/2dKdaMB

http://amzn.to/2eRuOjK

http://amzn.to/2eR5OTw

http://amzn.to/2frWHAz

http://amzn.to/2f14upz

http://amzn.to/2fPLsnR

http://amzn.to/2gzdgIz

http://amzn.to/2gBY3I0

Other books by Dean R. Giles

http://amzn.to/1KSZfgv

http://amzn.to/1UbXlSn

http://amzn.to/1LB1ub7

http://amzn.to/1LW8wrh